What Animal Am I?

by Caroline Hutchinson

I like acorns.

I get acorns from trees.

What am I?

I am a squirrel.

I can climb trees.

I like seeds.

I get seeds from sunflowers.

What am I?

I am a bird.

I can fly

to the sunflowers.

I like leaves.

I get leaves from trees.

What am I?

I am a deer.

I can eat leaves

from the trees.

I like flies.

I get flies in my web.

What am I?

I am a spider.